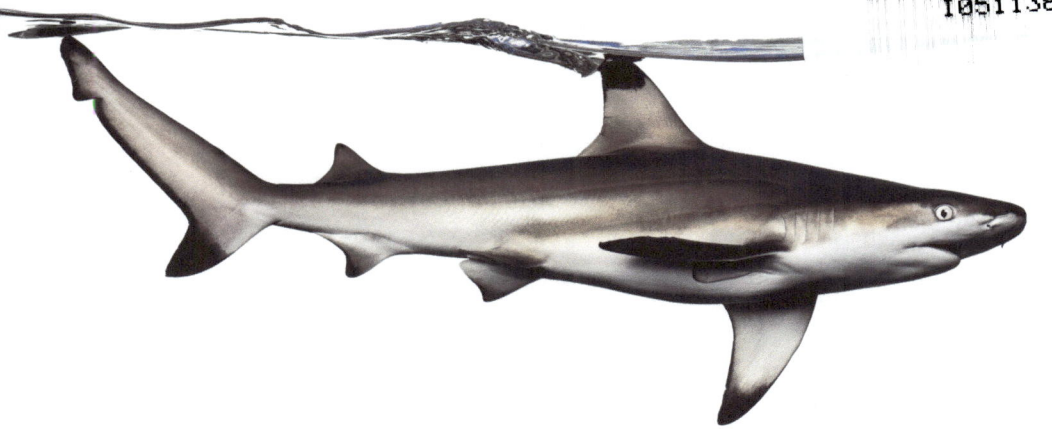

A Guidebook To
Blacktip Reef Sharks

Assembled by the Staff of Deep Sea Publishing

Edited by Eddie Hughes

Copyright Page

The sale of this book without a front cover is unauthorized. If this book is sold without a cover, then the author and the publishing company may have not received payment for it.

This book is a compilation of pictures and data from multiple sources. This book should not be considered a scientific journal and is more representative of a survey of information and photos of the Blacktip Reef Shark. All pictures were purchased royalty free or are public-domain, royalty-free images. References for all data has been provided. Editor for this work was Eddie R Hughes.

Copyright © 2020 by Deep Sea Publishing, LLC

All rights reserved. Published in the United States by

Deep Sea Publishing LLC, Bradenton, Florida.

Deep Sea Publishing ISBN-13: 978-1939535368

Deep Sea Publishing E-Book ISBN-13: 978-1939535375

www.deepseapublishing.com

Printed in the United States of America

eBook created in the United States of America

"Sharks have everything a scientist dreams of. They're beautiful—God, how beautiful they are! They're like an impossibly perfect piece of machinery. They're as graceful as any bird. They're as mysterious as any animal on earth. No one knows for sure how long they live or what impulses—except for hunger—they respond to. There are more than two hundred and fifty species of shark, and everyone is different from every other one."

— Peter Benchley, <u>Jaws</u>

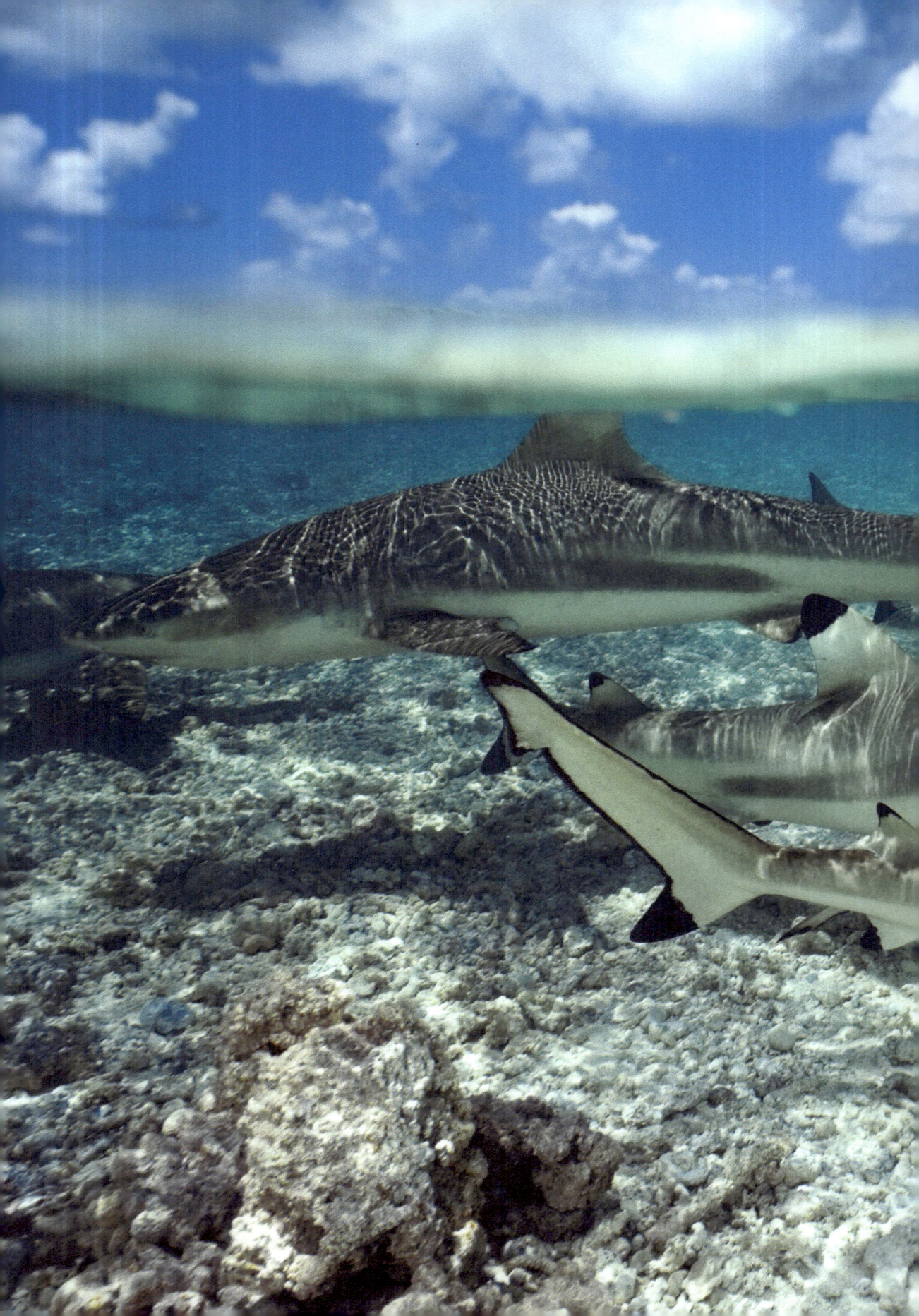

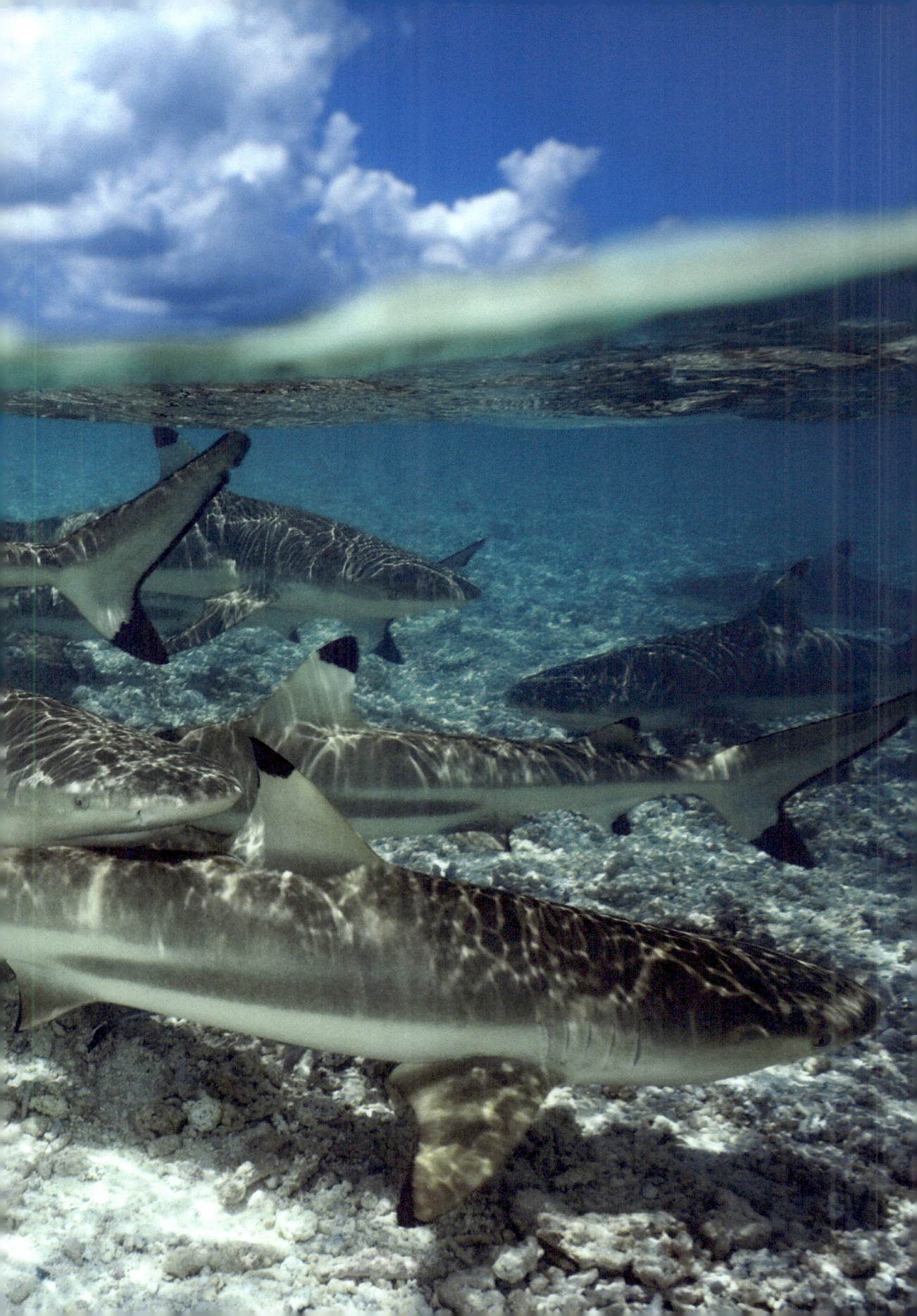

"Sharks are beautiful animals, and if you're luck enough to see lots of them, that means that you're in a healthy ocean. You should be afraid if you are in the ocean and you don't see sharks."

— Sylvia Earle

American marine biologist, author, explorer, lecturer

Carcharhinus melanopterus (Quoy & Gaimard, 1824)
Blacktip reef shark

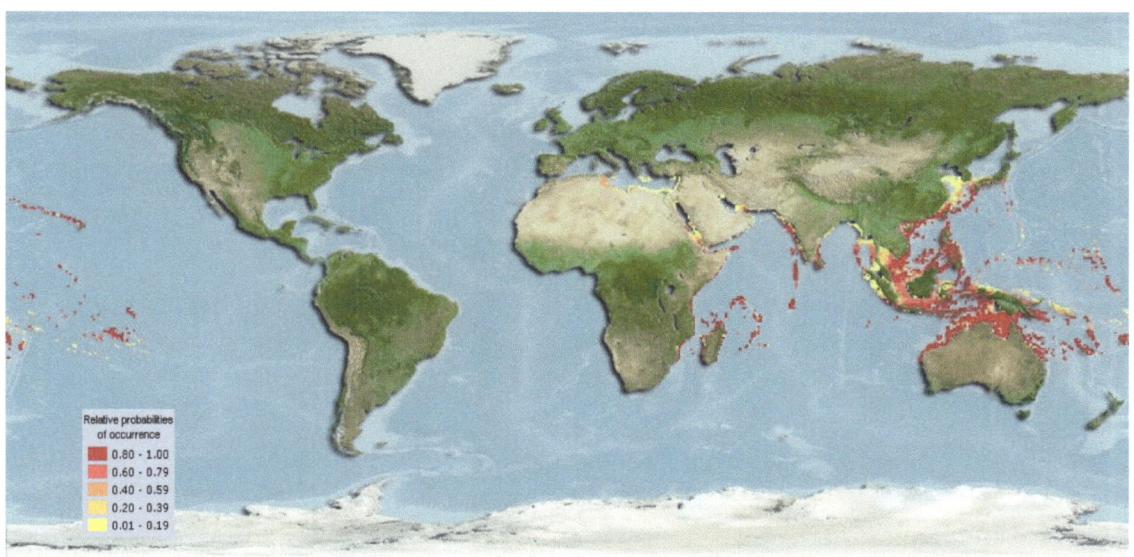

The Blacktip Reef Shark is likely to be seen at the locations are indicated on the map. Red is most likely; whereas, yellow is occasional.[1]

Kingdom:	Animalia
Phylum:	Chordata
Class:	Chondrichthyes
Subclass:	Elasmobranchii
Superorder:	Selachimorpha
Order:	Carcharhiniformes
Family:	Carcharhinidae
Genus:	Carcharhinus
Species:	C. melanopterus

Carcharhinus – animal with sharp nose. Derived from the Greek "karcharos" = sharp and "rhinos" = nose.

melano- – black

-pterous – wings

---

[1] Reviewed distribution maps for *Carcharhinus melanopterus* (Blacktip reef shark), www.aquamaps.org, version of Aug. 2013. Web. Accessed 23 Apr. 2016.

Kingdom: Animalia

Animalia contain organisms that are heterotrophic, multi-cellular, lack a cell wall, reproduce sexually, and are mobile at some point in their life.

Phylum: Chordata

Members of the phylum Chordata possess some unique characteristics that separate them from the other phylum. These include a notochord, and a nerve cord. A sub-phylum of this group is Vertebrates. Vertebrates possess a backbone, and contain organisms ranging from a African Forest Elephant (Loxodonta cyclotis) to a Ganges River Dolphin (Platanista gangetica).

Class: Elasmobranchii

This class, Elasmobranchii, contains sharks, skates, and rays. The key features that these organisms share are a skeleton made up of cartilage, rows of replaceable teeth, and the ability to feel small changes in electricity around them. They use this ability to detect changes in electricity to both find their way through the water as well as detect near by prey. Due to these characteristics, many of the organisms in this class are very skilled hunters, and in many cases, at the top of their food chain.

Order: Carcharhiniformes

Carcharhiniformes, or ground sharks, is the most abundant of the shark groups with about two-hundred known species. These species range from catsharks to hammerhead sharks usually dwelling in tropical areas. Ground sharks share common characteristics such as an anal fin, dorsal fins, and a nictitating membrane over the eye.

Family: Carcharhinidae

Members of Carcharhinidae lack spiracles (which is a respiratory opening found in many cartilaginous fish), very sharp blade like teeth, have a second dorsal fin that is smaller than the first, and a well developed nictitating membrane.

Genus: Carcharhinus

Members of the Carcharhinus include the Blacknose Shark, Copper Shark, and Whitetip Reef Shark (EOL 2013).

Species: Carcharhinus melanopterus

Carcharhinus melanopterus, better known as the Blacktip Reef Shark, is viviparous, meaning they produce live young, whereas many other sharks are oviparous meaning they lay eggs to be hatched outside the mother. Blacktip Reef Sharks are distinctively characterized by their black markings located on their fins. The markings are most dominant on the dorsal fin, and caudal fin, but are still visible on the pectoral, anal, pelvic, and secondary dorsal fin.[2]

---

[2] Campbell et all 208, *Biology*, Benjamin Cummings, San Francisco, California, USA.

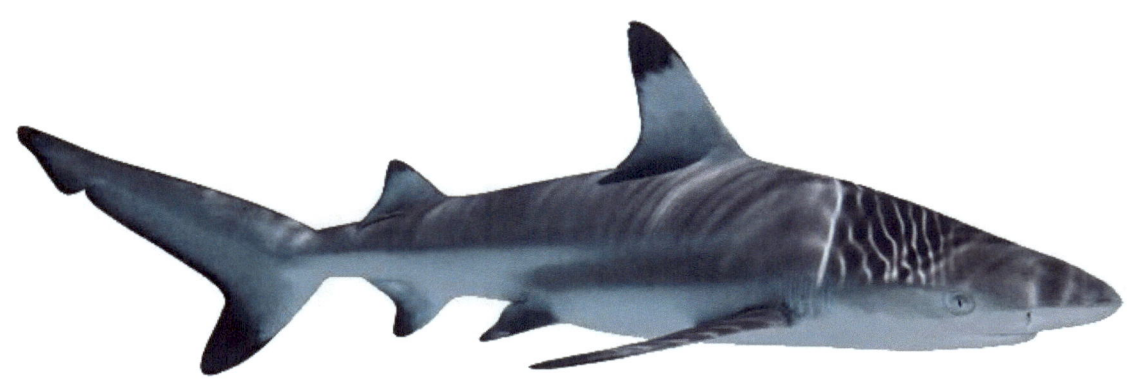

Species are classified by the IUCN Red List into nine groups,[15] set through criteria such as rate of decline, population size, area of geographic distribution, and degree of population and distribution fragmentation.

- Extinct (EX) – No known individuals remaining.
- Extinct in the wild (EW) – Known only to survive in captivity, or as a naturalized population outside its historic range.
- Critically endangered (CR) – Extremely high risk of extinction in the wild.
- Endangered (EN) – High risk of extinction in the wild.
- Vulnerable (VU) – High risk of endangerment in the wild.
- Near threatened (NT) – Likely to become endangered in the near future.
- Least concern (LC) – Lowest risk. Does not qualify for a more at-risk category. Widespread and abundant taxa are included in this category.
- Data deficient (DD) – Not enough data to make an assessment of its risk of extinction.
- Not evaluated (NE) – Has not yet been evaluated against the criteria.

When discussing the IUCN Red List, the official term "threatened" is a grouping of three categories: Critically Endangered, Endangered, and Vulnerable.

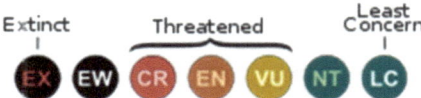

Blacktip reef shark (_Carcharhinus melanopterus_) is classified as near threatened. [3]

---

[3] The information on this page came from multiple sources, including the ICUN Red List of Threatened Species, Wikipedia, and the lower graphic is by Peter Halasz

Common Names for the Blacktip Reef Shark

English language common names for this species include blacktip reef shark, black tip reef shark, black fin reef shark, black finned shark, black tip shark, black tips nilow, black-tip reef shark, blackfin reef shark, blacktip shark, guliman, reef blacktip shark, and requien shark.

Other common names include aileron noir (Creole, French), anak hiu, hiu, ikan hiu, o chan, yu kepak hitam, yu nipah, yu shirip hitam, yu sirip hitam (Malay), apeape, malie-alamata (Samoan), bakebake (Gela), bako mij and pako (Marshallese), bakua, te bakoa and te baiburebure (Kiribati), balda, mori and khada mushi (Marathi), bayanakon (Banton), boka sorrah, bokka-sorrah, caval-sorrah, mukhan sorrah, nella vekal sorrah, raman sorrah, ran sorrah (Telugu), cá mâp vây den, (Vietnamese), chalarm hoo-dum (Thai), falhu miyaru (Maldivian), fungu, marracho tinteiro de coral, nyatussue, tabar, tubarão, tubarão and xituo (Portuguese), gunna sura, kalakumattai-sura, katta-sura, thalan-sorrah and koppulisura (Tamil), gursh, jahrah, jarjur, quash asswad, rabie and shattafi (Arabic), iho and pating (Waray-waray), kaitan tutungan and pating (Maranao/Samal/Tao Sug), lodlod and tutongan (Bikol), lumba, pating, pating inglesa and pantay (Tagalog), ma'o mauri, requin à pointes noires, requin noir and requin pointes noires (French), magara and mossikhada (Gujarati), mago (Niuean), manô pâ'ele (Hawaiian), mauri (tahitian), mookan-sravu (Malayalam), mustaevähai (Finnish), neikaplethantee (Kannada), nga-man-taung-me (Burmese), papa karaji (Swahili), pating (Magindanaon), peshkaqen (Albanian), peu and woshaalang (Carolinian), sorrippet revhaj (Danish), svartspetshaj (Swedish), swartvin-rifhaai (Afrikaans), te baiburebure (Tuvaluan), teburon (Kuyunon), tiburón de puntas negras (Spanish), tsuma guro and tsumaguro (Japanese), Žralok cernošpicí and Žralok útesový cernošpicý (Czech) and Zwartpuntrifhaai (Dutch).[4]

---

[4] Florida Museum of Natural History: http://www.flmnh.ufl.edu/fish/discover/species-profiles/carcharhinus-melanopterus

"Many people continue to think of sharks as man-eating beasts. Sharks are enormously powerful and wild creatures, but you're more likely to be killed by your kitchen toaster than a shark!"

— Ted Danson

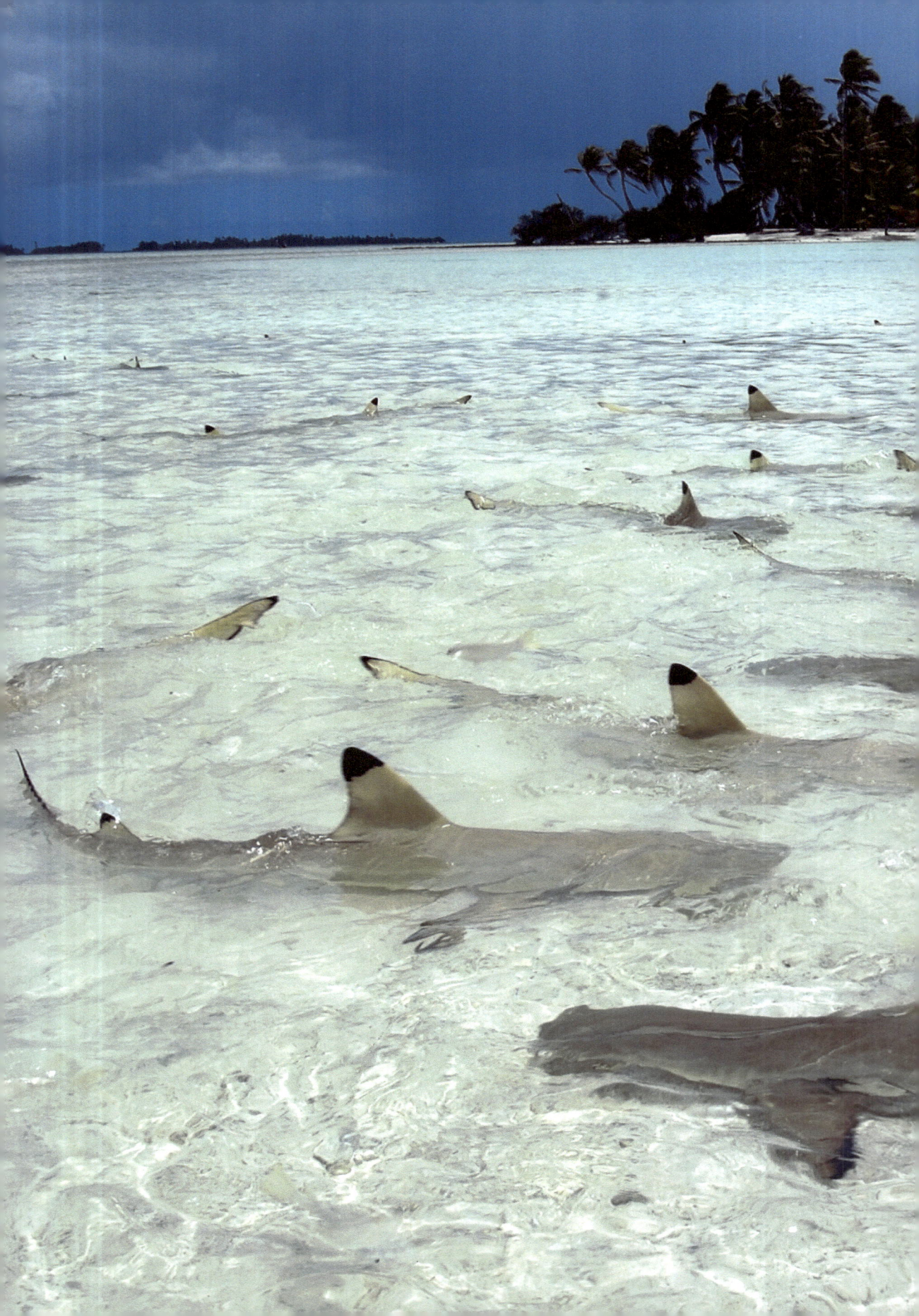

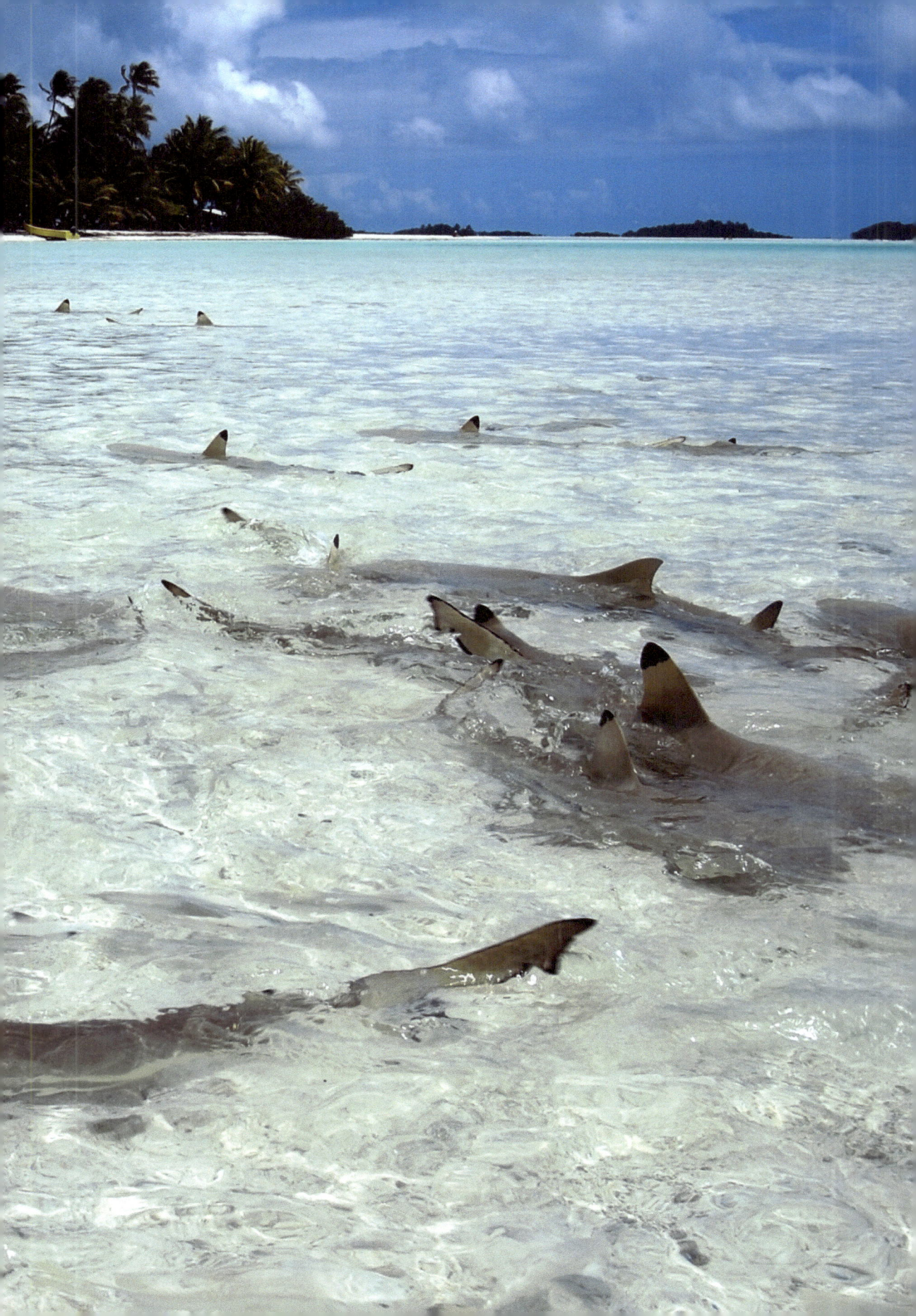

'I was swallowed by a shark, a great big ugly shark', 'Eek!, Eek!, Sharks!', goes a children's song. Songs like these and millennia of damaging publicity have given sharks a bad name, so that now, aversion and fear is the default reaction to sharks, if not craving for their fins.

— Angelique M. Songco
Protected Area Supervisor
Tubbataha Reefs Natural Park, Philippines
Sharks and Rays of Tubbataha Reef

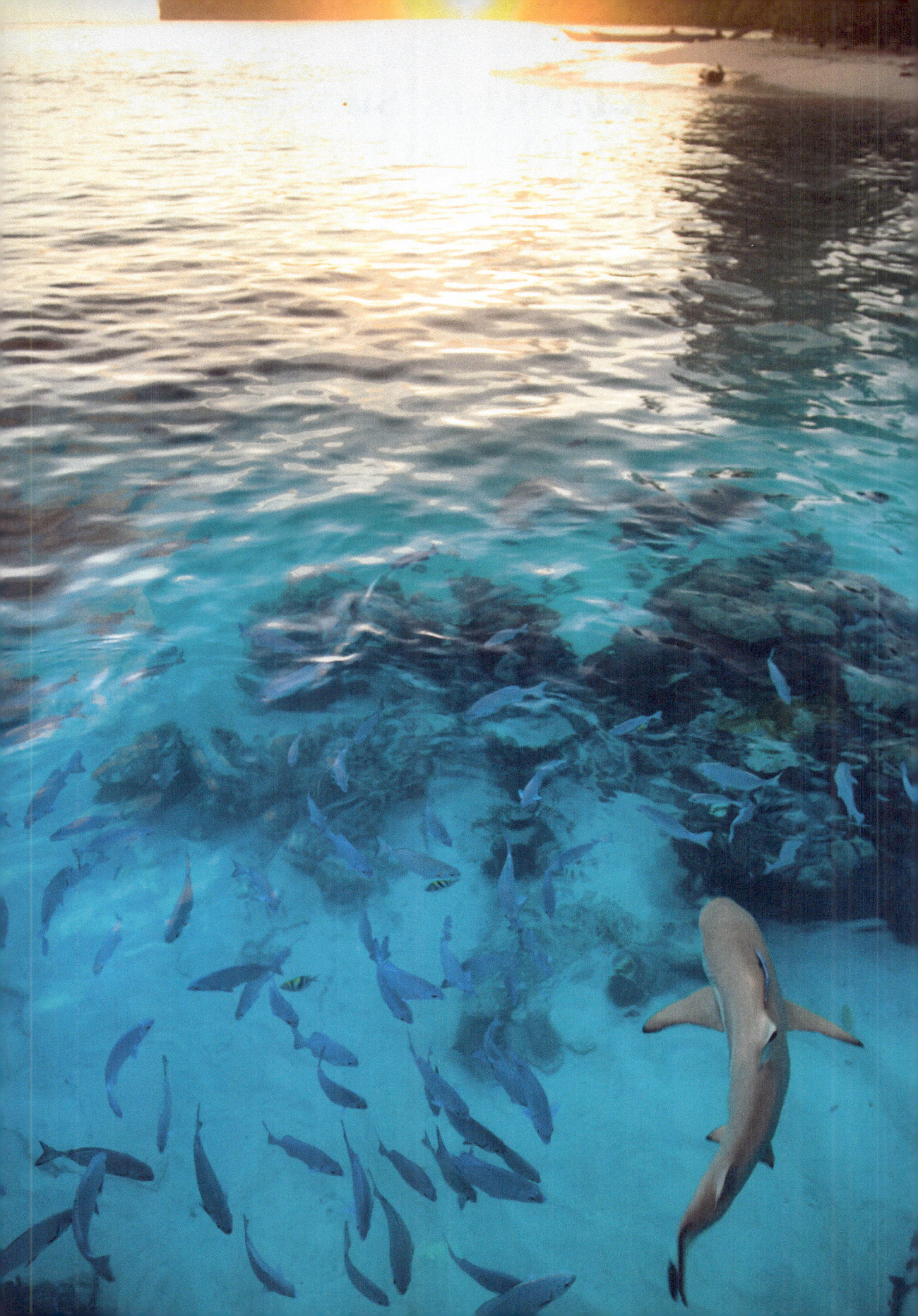

BLACKTIP REEF SHARKS ARE RELATED TO BLACKTIP SHARKS

The blacktip reef shark belongs to a family of 54 shark species called requiem sharks. Some of the best-known and most abundant sharks are requiem sharks. The word "requiem" comes from the French word for shark, "requin."

Many requiem sharks are swift but aggressive hunters. Tiger, bull, and oceanic whitetip sharks are all requiem sharks. These excellent swimmers are found in a huge variety of habitats, including open oceans and shallow reefs.

The blacktip reef shark (*Carcharhinus melanopterus*) is often confused for its close cousin, the blacktip shark (*Carcharhinus limbatus*). While the blacktip reef shark is grayish brown, blacktip sharks are gray or blue in color. The blacktip reef shark has a distinct black triangle at the tip of its dorsal fin, while only the trailing edge of a blacktip shark's dorsal fin is black. Also, blacktip sharks usually don't have black tips on their anal or pelvic fins.

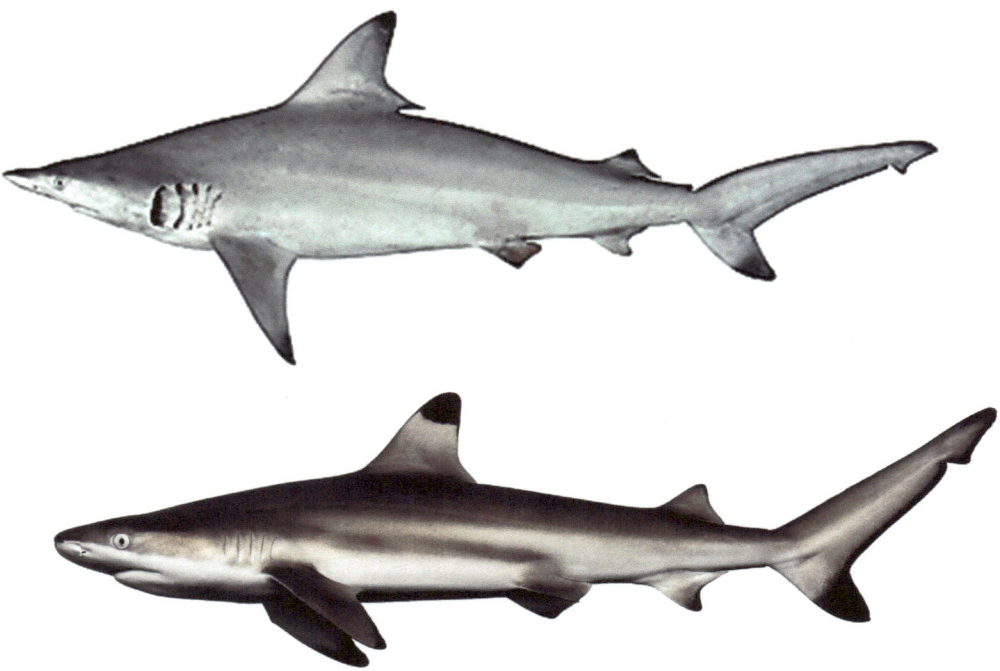

Location can also set these two sharks apart. Blacktip reef sharks are found more commonly in the Indo-Pacific, while blacktip sharks are often found in the Western Atlantic Ocean. So if you see a shark with black tips while diving in Florida it is a blacktip shark, not a blacktip reef shark.[5]

---

[5] The information on this page was from a variety of public domain sources.

"Sharks don't eat humans," shark exert Peter Kimley of the University of California, Davis told LiveScience. "They spit out humans. Humans aren't nutritious enough to be worth the effort."

http://www.nbcnews.com/id/8473619/ns/technology_and_science-science/t/science-shark-attacks/#.VxrLoPkrKM8

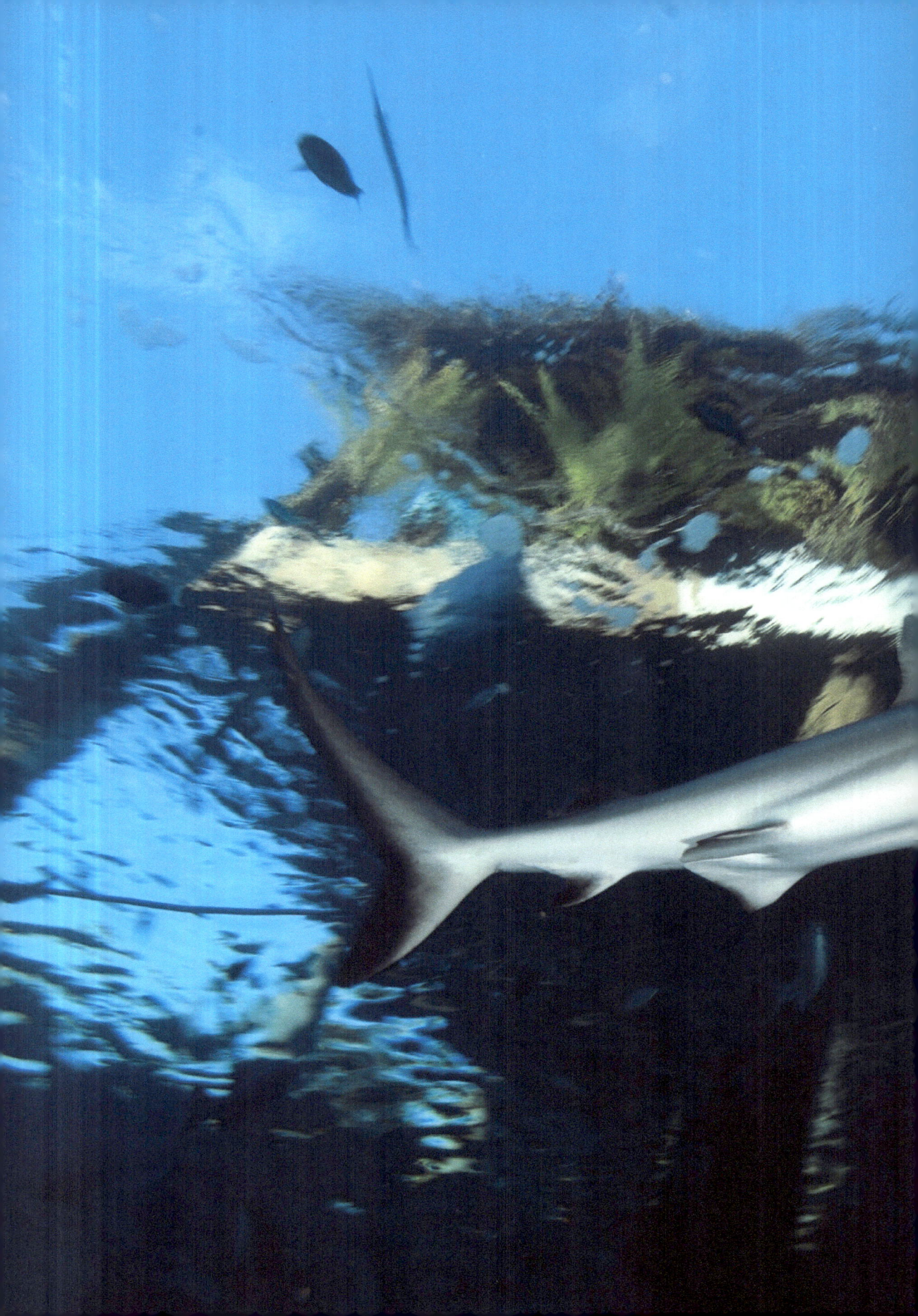

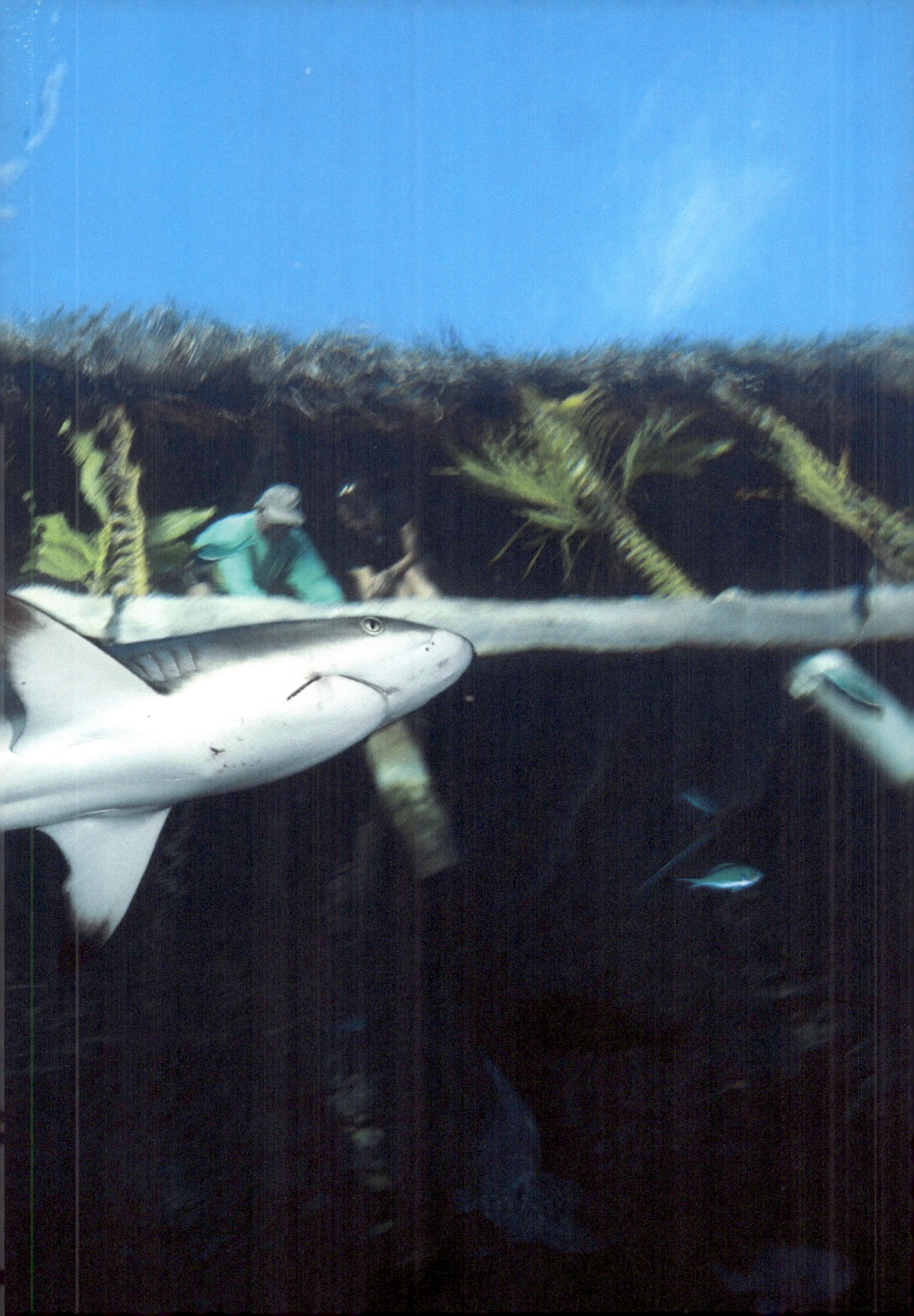

"Considering their impact, you might expect mosquitoes to get more attention than they do. Sharks kill fewer than a dozen people every year, and in the U.S. they get a week dedicated to them on TV every year."

— Bill Gates

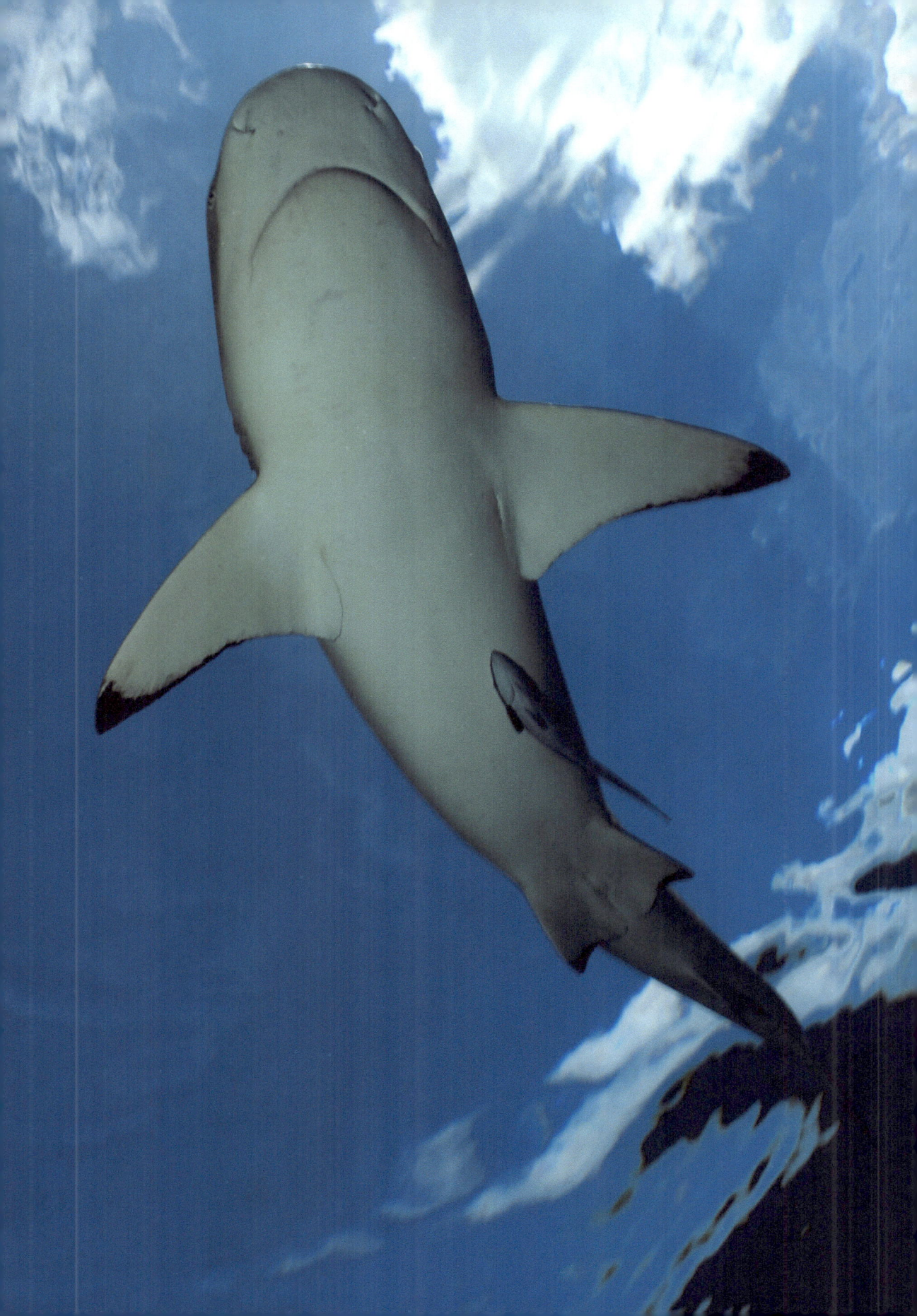

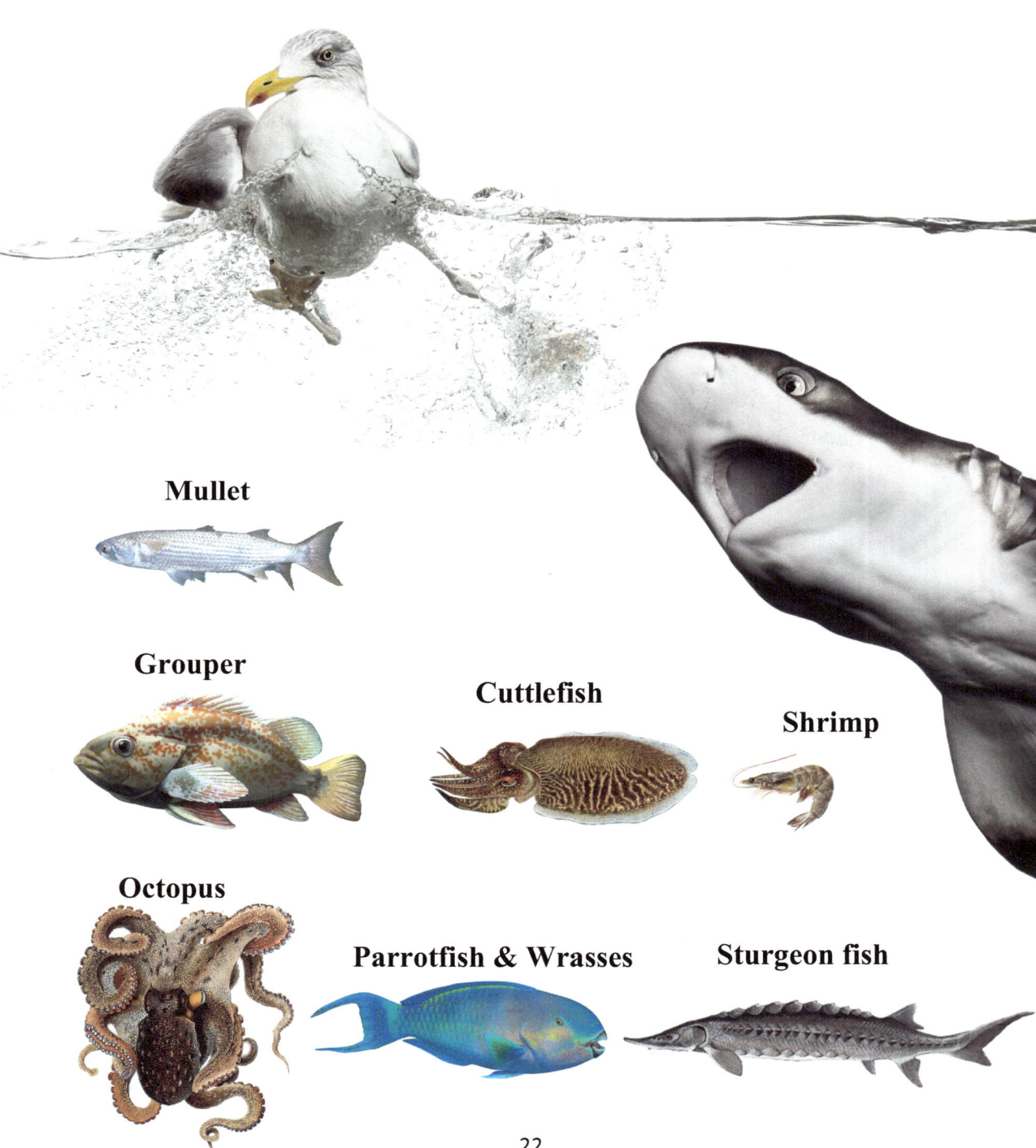

Food

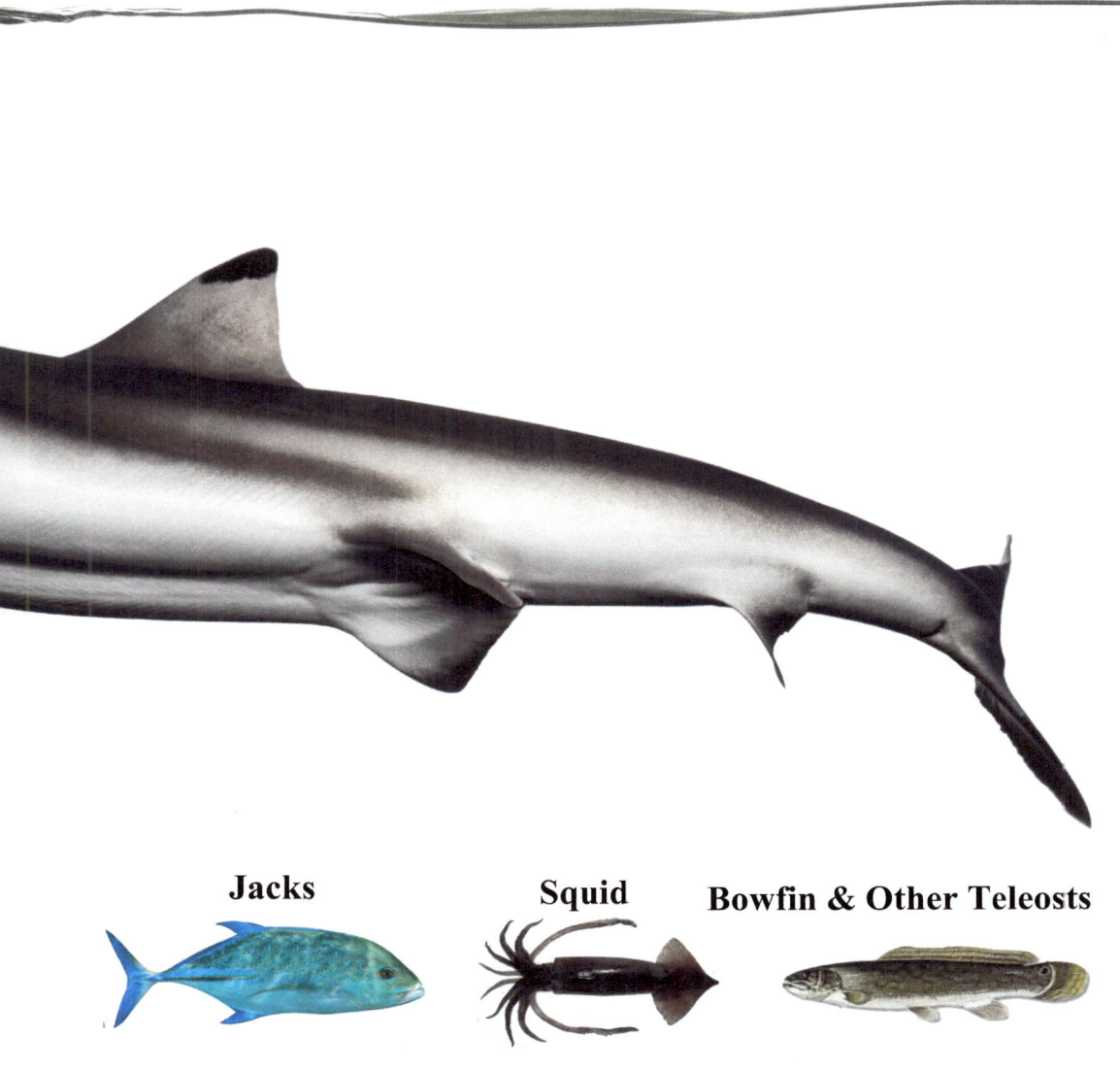

Jacks **Squid** **Bowfin & Other Teleosts**

Anatomy

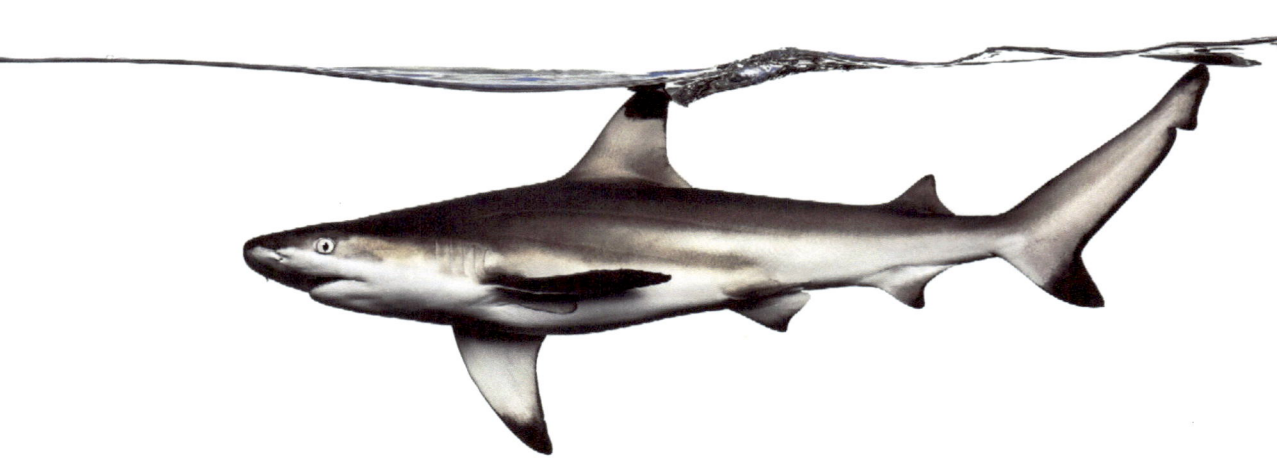

Blacktip Reef Shark
Carcharhinus melanopterus

Description
- 1st dorsal-fin tip sharply defined, black highlighted beneath by white
- All fins with conspicuous black tips
- Interdorsal ridge absent
- Snout is very short, broadly rounded (viewed from underneath), preoral length subequal to internarial space
- Upper teeth with narrow, oblique central cusp and low basal cusplets
- Lower teeth narrow, upright to oblique, edges finely serrated

Size
Attains at least 140 cm, possibly to 180 cm; males mature at 98–113 cm and females at 96–120 cm; born at 48–50 cm.[6]

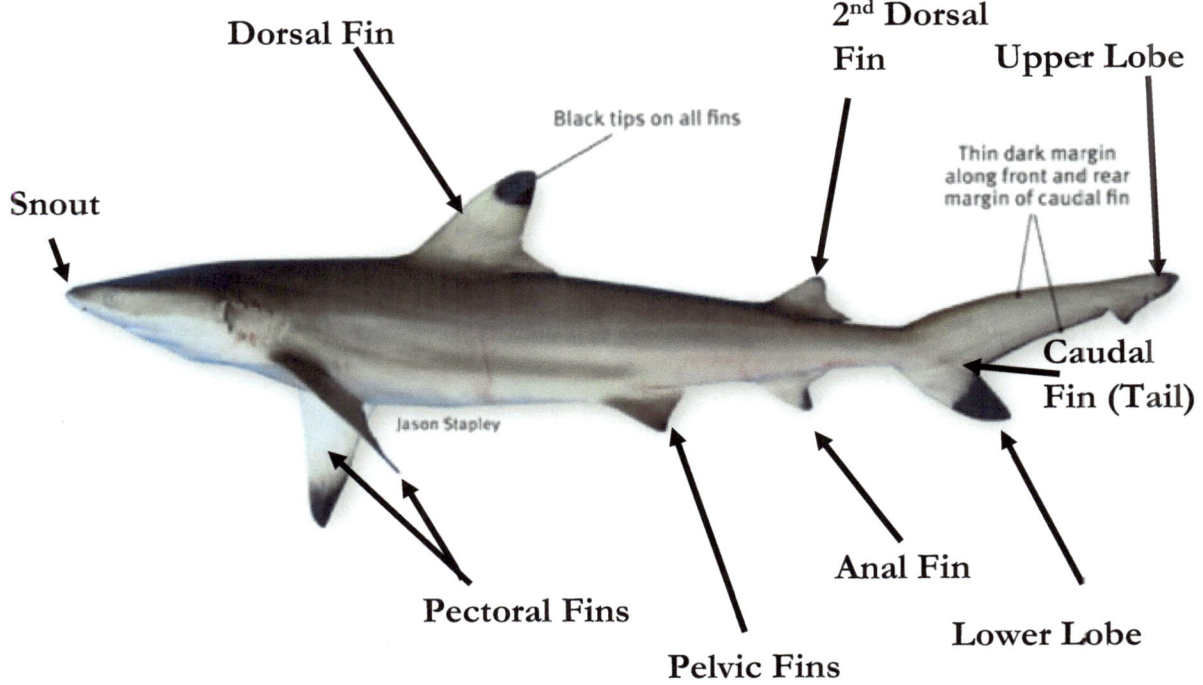

---

[6] h:tp://tubbatahareefs.org/wp-content/uploads/2012/11/Sharks-and-Rays-of-Tubbataha-Reef.pdf

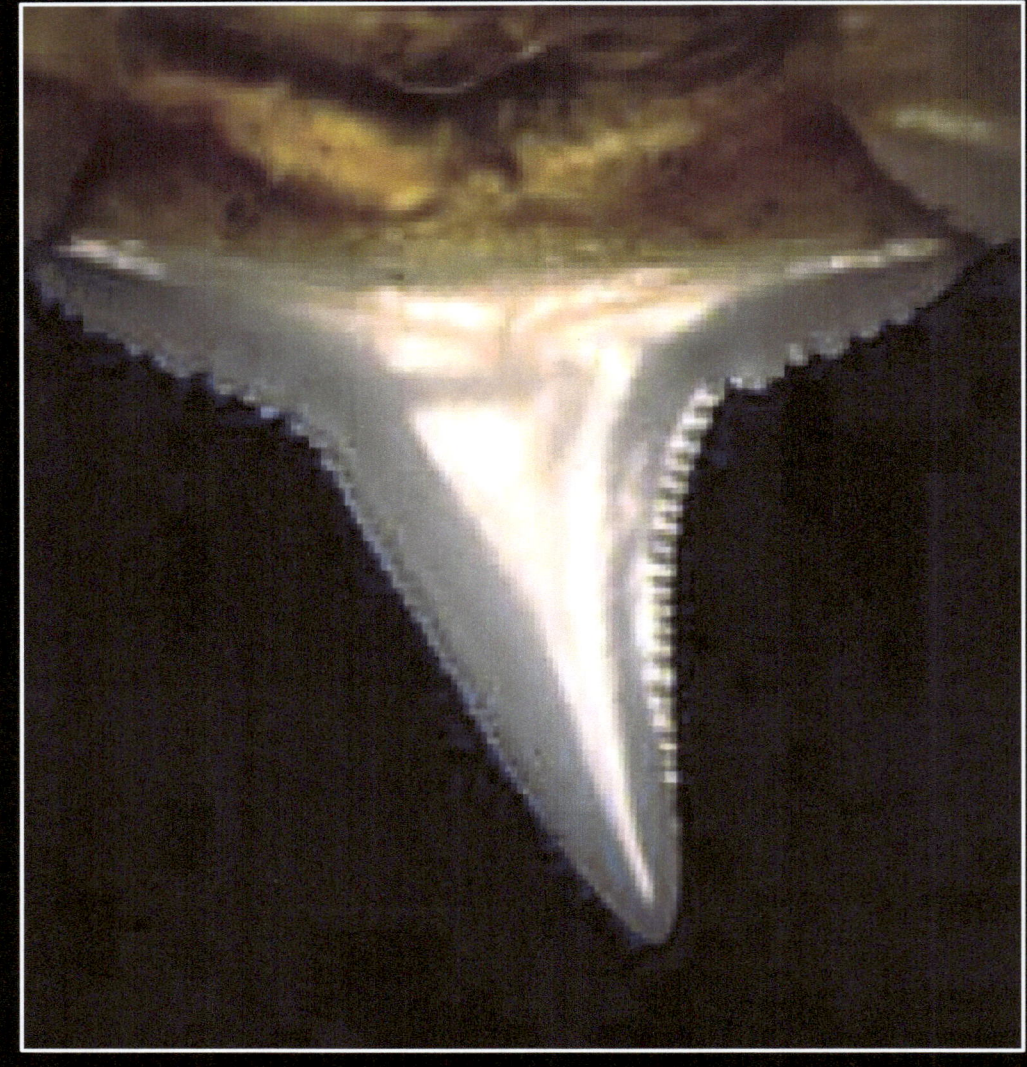

The upper teeth of the blacktip reef shark are narrow, semi-oblique and serrated with notched margins at the base.

Picture and description from:

https://www.daf.qld.gov.au/fisheries/species-identification/shark-identification-guide/photo-guide-to-sharks/sharks,-part-1/blacktip-reef-shark

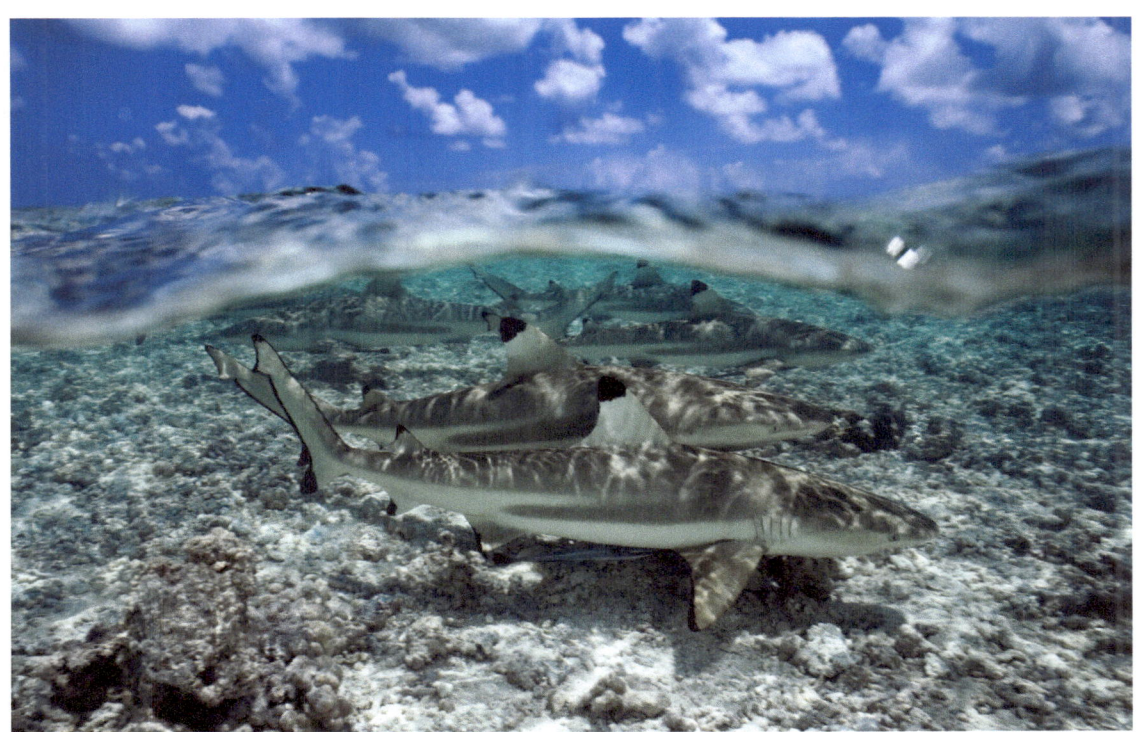

Reproduction

Reproduction is viviparous (the embryo develops inside the body of the mother, as opposed to outside in an egg), with 2 to 4 pups in a litter. Before giving birth, female Blacktip Reef sharks will incubate their young for 10-11 months in the Indian Ocean and 7-9 months in the Pacific Ocean.[7] The pups' length at birth ranges from 33 to 52 centimetres.[8] Typically the larger pups are found in the Indian Ocean and off northern Australia.

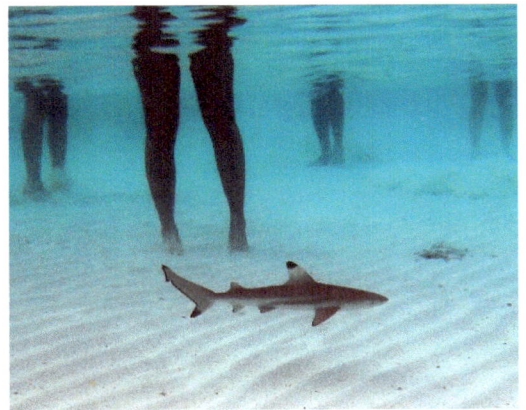

Baby Blacktip Reef Shark (right)[9]

---

[7] Porcher, I.F. (April 2005). "On the gestation period of the blackfin reef shark, Carcharhinus melanopterus, in waters off Moorea, French Polynesia". Marine Biology. 146 (6): 1207–1211.
[8] https://animalcorner.org/animals/blacktip-reef-shark/
[9] https://www.pinterest.com/pin/408631366160897894/

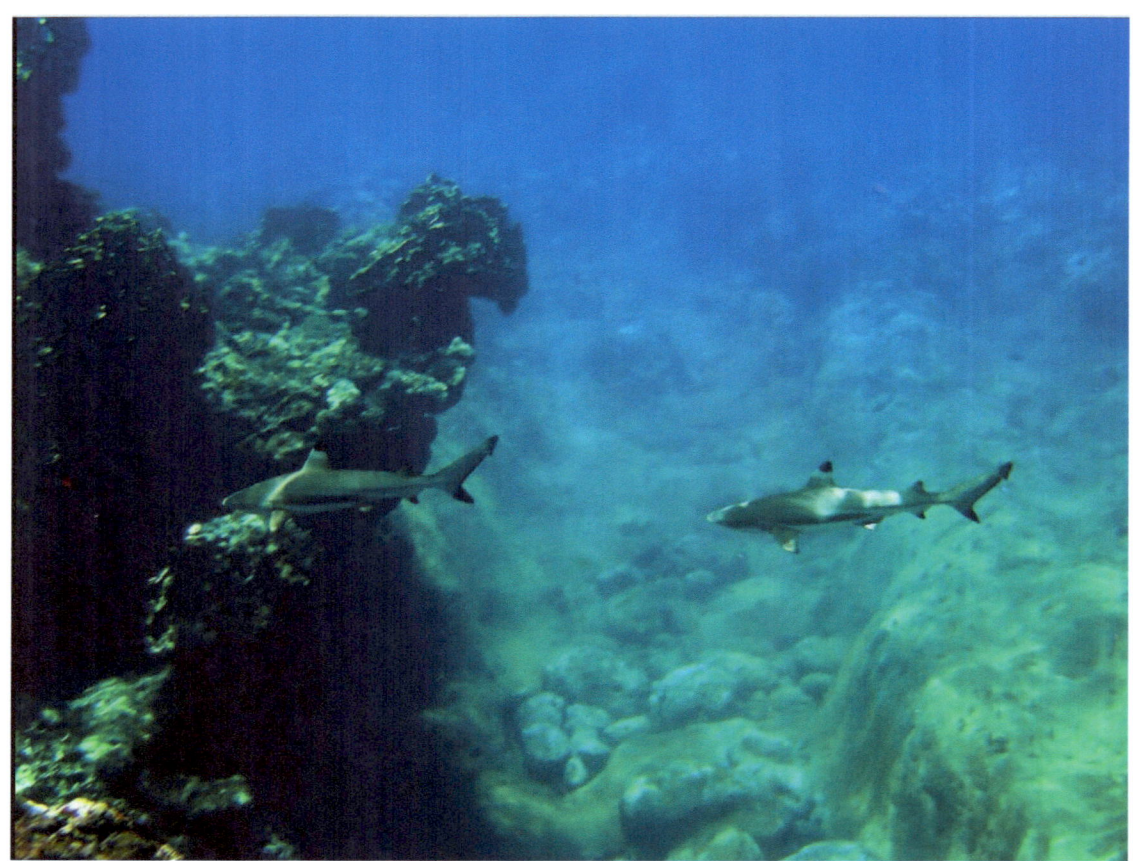

Blacktip Reef Sharks will follow each other as a prelude to mating as shown in the picture above.[10] These two sharks are swimming off the coast of Guam. The male with often bite the female on her pectoral fins, which takes about 4-6 weeks to heal.[11] Mating takes place on sandy bottoms and normally takes a few minutes to complete.[12]

Interestingly enough, females are able to reproduce asexually if males are not available.[13]

---

[10] David Burdick, NOAA - http://www.photolib.noaa.gov/htmls/reef0977.htm.   Public Domain.
[11] See footnote 6.
[12] Johnson, R.H. & D.R. Nelson (1978). "Copulation and possible olfaction-mediated pair formation in two species of carcharhinid sharks". Copeia. American Society of Ichthyologists and Herpetologists. 1978
[13] https://www.sharksider.com/blacktip-reef-shark/

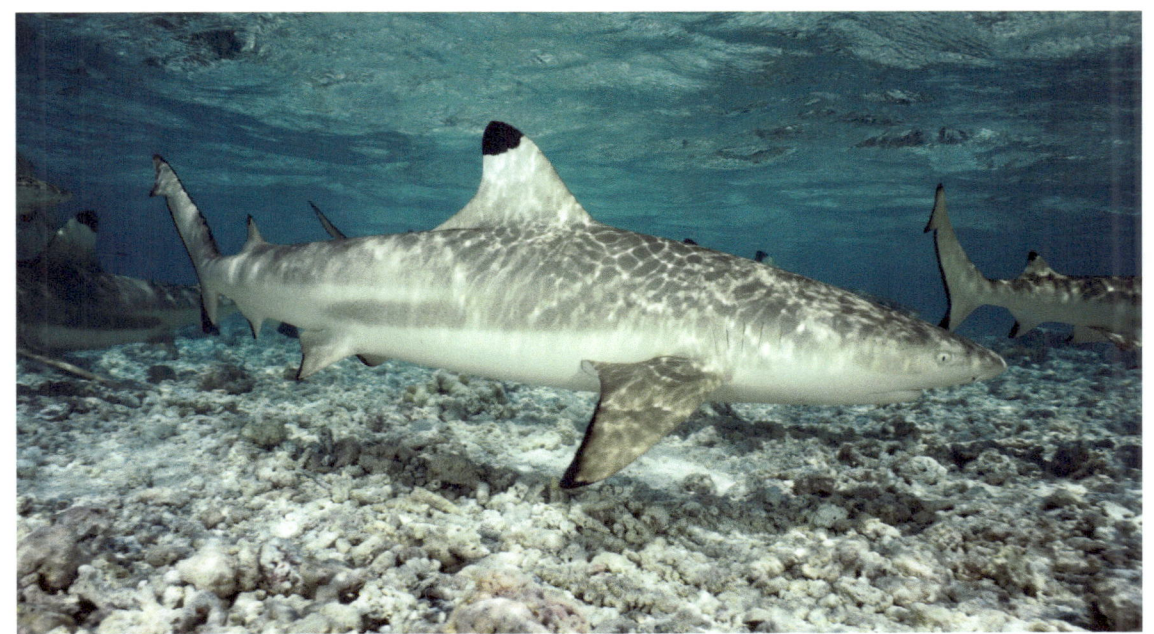

Interactions with Humans

The Blacktip Reef Shark is not considered social, however, it can be seen in small groups. While generally shy, they are often curious about snorkelers and scuba divers. As with most sharks, the body is bent into a sort of 'S' shape when the shark feels threatened. Blacktip Reef sharks are harmless unless provoked. Incidents generally involve hand feeding or spear fishing, possibly in combination with low visibility.[14] The International Shark Attack File (ISAF) has recorded just 11 unprovoked blacktip reef shark bites on humans since 1959.[15]

Blacktip reef sharks are not listed as threatened or endangered anywhere in the world but populations are declining, mostly due to overfishing. Commercial fisheries often target this species for their meat, liver oil, and fins. Their small litter sizes and longer gestation periods make them especially vulnerable to population decline since it takes longer to replace individuals removed from the population.[16]

---

[14] https://animalcorner.org/animals/blacktip-reef-shark/
[15] https://www.floridamuseum.ufl.edu/discover-fish/species-profiles/carcharhinus-me anopterus/#:~:text=Blacktip%20reef%20shark%20have%20occasionally,bites%20on%20humans%20since%201959.
[16] https://marinesanctuary.org/blog/sea-wonder-blacktip-reef-shark/?gclid=EAIaIQobChMIx_iZ59-T6vIVGK_ICh33qwwgEAAYASAAEgI01_D_BwE

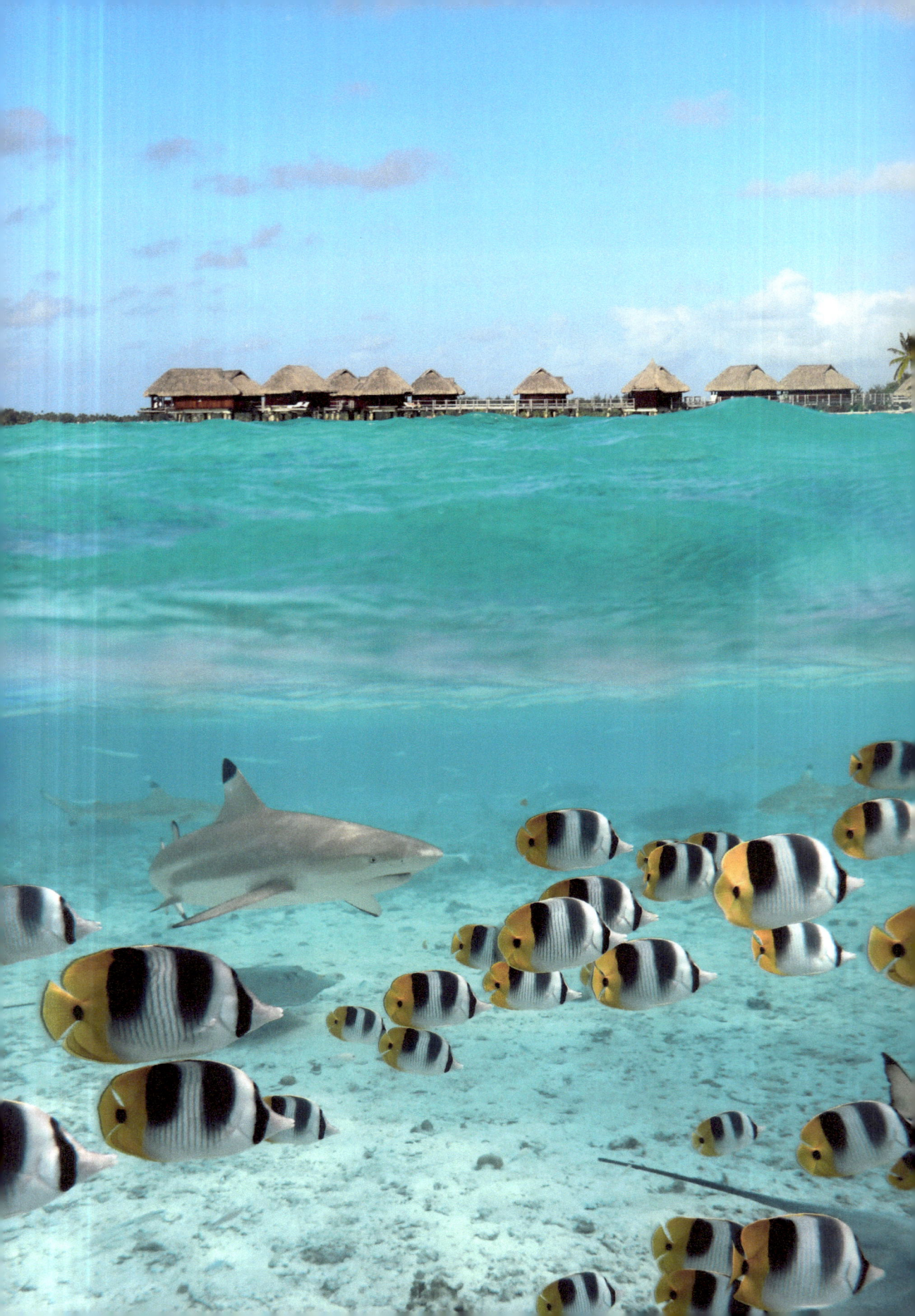

"Sharks are as tough as those football fans who take their shirts off during game in Chicago in January, only more intelligent."

— David Barry
Americal author and was a columnist who wrote a nationally syndicated humor column.

Blacktip Reef Shark Crossword Puzzle

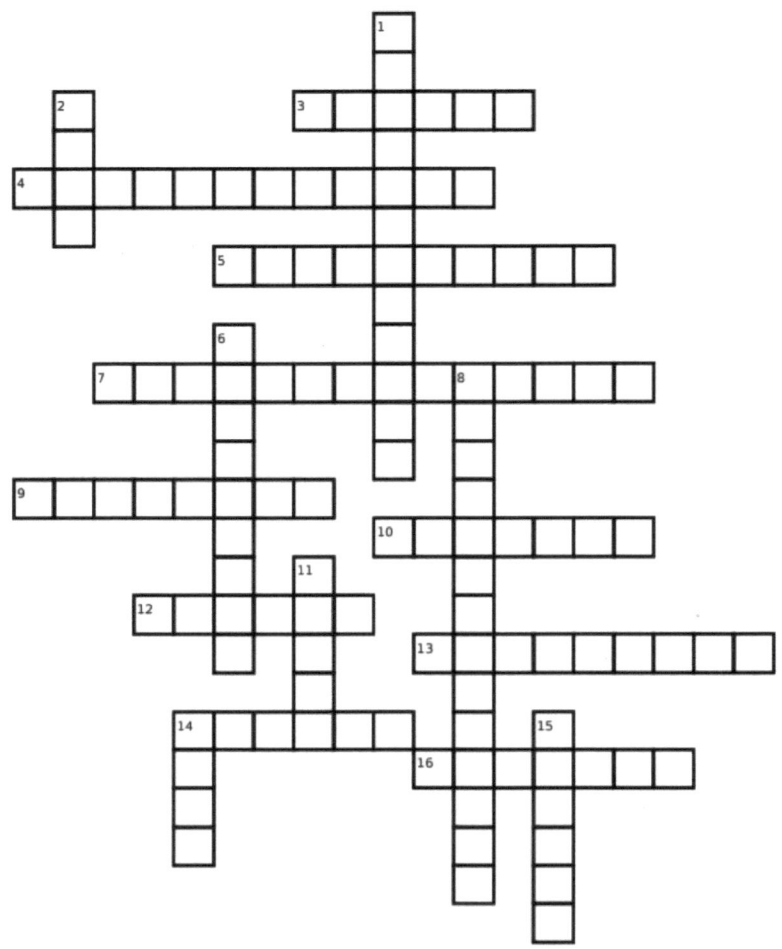

Down:	Across:
1. Species of the Blacktip Reef Shark.	3. Fins found on the bottom of this shark.
2. Has recorded just 11 unprovoked Blacktip reef shark bites on humans since 1959.	4. Genus of Blacktip Reef Shark.
6. Japanese word for Blacktip Reef Shark.	5 The embryo develops inside the body of the mother, as opposed to outside in an egg.
8 (2 words) Likely to become endangered in the near future.	7. Class of Blacktip Reef Shark
11. Tahitian word for Blacktip Reef Shark	9. Phylum of Blacktip Reef Shark.
14. The group that generates the Red List of endangered species.	10. The Blacktip Reef Shark likes to eat this soft-bodied, eight-limbed mollusk.
15. A food choice of the Blacktip that is also consumed by humans – often smoked and in dips.	12. This is the upper fin of the Blacktip.
	13. The state of the Blacktip population.
	14. One of two oceans where Blacktips live.
	16. The Blacktip Reef Shark is a member of this family of 54 sharks.

References

Pages 4-5 picture: Yann Hubert | Dreamstime.com - <a href="http://www.dreamstime.com/stock-photo-blacktip-reef-shark-carcharhinus-melanopterus-half-underwater-view-swimming-image60655168#res3171049">BLACKTIP REEF SHARK/CARCHARHINUS MELANOPTÉRUS</a>

Page 27 picture: Yann Hubert | Dreamstime.com - <a href="http://www.dreamstime.com/stock-photo-blacktip-reef-shark-half-underwater-view-swimming-image60655358#res3171049">BLACKTIP REEF SHARK</a>

Page 23 picture: Yann Hubert | Dreamstime.com - <a href="http://www.dreamstime.com/stock-photo-blacktip-reef-shark-swimming-surface-clear-water-image60784476#res3171049">BLACKTIP REEF SHARK</a>

Pages 18-19 Picture: Yann Hubert | Dreamstime.com - <a href="http://www.dreamstime.com/stock-photo-blacktip-reef-shark-swimming-under-over-water-house-image60784076#res3171049">BLACKTIP REEF SHARK</a>

Page 15 Picture: Streliuk Aleksei | Dreamstime.com - <a href="http://www.dreamstime.com/stock-photo-shark-pure-water-sunset-papua-image62402220#res3171049">Shark in pure water at sunset</a>

Page 29 Picture: Pljvv | Dreamstime.com - <a href="http://www.dreamstime.com/stock-photo-shark-butterfly-fish-bora-bora-image25913630#res3171049">Shark and butterfly fish at Bora Bora</a>

Pages 22-23 Picture: Isselee | Dreamstime.com - <a href="http://www.dreamstime.com/royalty-free-stock-image-shark-hunting-gull-blacktip-reef-european-herring-image39255366#res3171049">Shark hunting a Gull</a>

Pages 1- 24 Picture: Isselee | Dreamstime.com - <a href="http://www.dreamstime.com/royalty-free-stock-photos-blacktip-reef-shark-swimming-surface-side-view-water-carcharhinus-melanopterus-isolated-white-image38841018#res3171049">Blacktip reef shark swimming at the surface</a>

Most quotes from famous people listed in this book are found on brainyquote.com.

Front Cover picture: Yann Hubert | Dreamstime.com <a href="https://www.dreamstime.com/ yannhubert_info" itemprop="author">Yann Hubert</a>

Deep Sea Publishing has made every effort to document the sources of information in this book. Permission to use photos that were not public domain were properly obtained and the require references are listed either below the picture or on this page.

Answers to the Crossword puzzle is found on www.deepseapublishing.com. Crossword created on Education.com.

www.ingramcontent.com/pod-product-compliance
Lightning Source LLC
Chambersburg PA
CBHW041935240526
45473CB00034B/1704